Cambridge
checkpoint

Cambridge Assessment
International Education
Endorsed for learner support

Lower Secondary
Science

WORKBOOK

9

Peter D Riley

Boost

HODDER
EDUCATION

Acknowledgements

The Publishers would like to thank the following for permission to reproduce copyright material.

Cambridge International copyright material in this publication is reproduced under licence and remains the intellectual property of Cambridge Assessment International Education.

The Boost knowledge tests and answers have been written by the authors. These may not fully reflect the approach of Cambridge Assessment International Education.

Photo credits

p.10 *l* © Michael Ireland - Fotolia; **p.10** *c* © Lizziemaher/stock.adobe.com; **p.10** *r* © Melinda Fawver/stock.adobe.com; **p.12** © Igor Dudchak/stock.adobe.com; **p.13** © Gelpi/stock.adobe.com; **p.15** © Iredding01/stock.adobe.com; **p.24** © Rafael Ben-Ari/stock.adobe.com; **p.75** *t, l* © De Agostini Picture Library/Getty Images; **p.75** *t, r* © David Costa Art/stock.adobe.com; **p.75** *b* © Catmando/stock.adobe.com

Every effort has been made to trace all copyright holders, but if any have been inadvertently overlooked, the Publishers will be pleased to make the necessary arrangements at the first opportunity.

Although every effort has been made to ensure that website addresses are correct at time of going to press, Hodder Education cannot be held responsible for the content of any website mentioned in this book. It is sometimes possible to find a relocated web page by typing in the address of the home page for a website in the URL window of your browser.

Third-party websites and resources referred to in this publication have not been endorsed by Cambridge Assessment International Education.

Hachette UK's policy is to use papers that are natural, renewable and recyclable products and made from wood grown in well-managed forests and other controlled sources. The logging and manufacturing processes are expected to conform to the environmental regulations of the country of origin.

Orders: please contact Hachette UK Distribution, Hely Hutchinson Centre, Milton Road, Didcot, Oxfordshire, OX11 7HH. Telephone: +44 (0)1235 827827. Email: education@hachette.co.uk Lines are open from 9 a.m. to 5 p.m., Monday to Friday.

You can also order through our website: www.hoddereducation.com

ISBN: 978 1 3983 01436

© Peter D Riley Ltd 2022

First published in 2005

Second edition published in 2011

This edition published in 2022 by
Hodder Education,
An Hachette UK Company
Carmelite House
50 Victoria Embankment
London EC4Y 0DZ

www.hoddereducation.com

Impression number 10 9 8 7 6 5 4 3

Year 2026 2025 2024 2023

Cover photo © NicoElNino - stock.adobe.com

Illustrations by Integra Software Services

Typeset by Integra Software Services Pvt. Ltd., Pondicherry, India

Printed in the UK

A catalogue record for this title is available from the British Library.

Contents

Introduction

The aim of every science course is to help you become scientifically literate or, more simply, to help you become a 'scientific citizen'. This means that you can confidently talk and write about the science you have studied and know how it helps us to understand and live in our world. Below are some questions from this book that a scientific citizen should be able to answer. Just read through them slowly.

What does the word 'transpiration' mean? How does the kidney remove urea from your blood? What does a plant need for photosynthesis? What does it produce in the process of photosynthesis? What is used in hospitals to help premature babies survive? What is an endangered species? All the elements known to science are set out in a table; what is it called? What do atoms form to join themselves together? What is meant by the density of a substance? What is temperature? What is the amplitude of a wave? In a circuit diagram, what is the symbol for a cell and what is a symbol for a lamp? What is a seismometer for? What is a greenhouse gas? Where is the asteroid belt?

The chances are that you will not be able to answer most of those questions now, but if you work through this book as you complete the chapters in the *Checkpoint Science Stage 9 Student's Book*, you will be on your way to being scientifically literate.

Here is the first challenge of the book – look at these questions again and write down any answers you might have. It does not matter if you cannot think of any answers to a question, just keep a record of your answers for later. When you have completed this workbook, look back at these questions again, write down your answers and see how they differ from your first answers. This should show you that you are well on your way to being a scientific citizen.

Most of the questions aim to test your knowledge and understanding of science, but some questions have this icon ⭐. These questions aim to test your science enquiry skills.

Some other questions have this icon ⬡. These are questions about using models to help you learn about and understand scientific ideas.

The questions that have this icon 🔗 require you to use skills that you are learning in other subjects that you are studying.

Yet other questions have this icon ⊞. These questions put science in context.

So now it is time to start. Read each question carefully, think about it, then write down the answer in the space provided, or as otherwise instructed. Often you will have to write down some facts or explanations, sometimes you will need to tick a box and occasionally you will have to link details by lines or construct and interpret graphs.

1 Water and life

The transport of water through the root and root hairs

1 What is the main function of root hairs in a plant?

....

Transport of water through the stem and leaves

 2 Students have set up an experiment to observe how water passes up the stem in six celery plants.

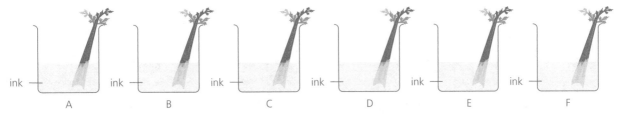

| ink | ink | ink | ink | ink | ink |
| A | B | C | D | E | F |

- Each celery stem is 15 cm tall.
- They put each celery stem in the orange ink at the same time.
- They take the stems out of the water at 10-minute intervals.
- They cut open the stem in slices to see how far the water has reached up each stem.

The table below shows their results.

Beaker	Time/minutes	Height of water/cm
A	0	1.0
B	10	1.7
C	20	2.9
D	30	4.0
E	40	3.2
F	50	7.0

a Draw a graph of the results in the space below. Label the *x*-axis 'time in minutes' and the *y*-axis 'height of water in cm'.

b One of these results is an anomalous result. **Circle** this result on your graph.

c How can you tell this is the anomalous result? Give a reason why it might be anomalous.

....

....

d Complete this sentence to make your conclusion:

The .... the time, the .... the water rose up the stem.

e The information that you plotted on your graph is known as data points. Use them to estimate how many minutes it might take the orange dye to reach the celery leaves.

.... minutes.

3 What does the term 'transpiration stream' mean?

....

....

Minerals and how plants use them

4 Two important minerals for plants are magnesium and nitrates. Draw a line from the name of each mineral to link it with the main use it has in a plant.

| magnesium |

| makes food |

| makes protein |

| nitrates |

| anchors the plant |

| makes chlorophyll |

| takes up water |

The human renal system

5 Which statements about the human renal system are correct? Tick (✓) **two** boxes.

The kidneys filter the blood. ☐

The kidneys remove urea in urine. ☐

The kidneys are near your heart. ☐

The kidneys make blood cells. ☐

6 Which pieces of equipment could you use to model how the kidneys process blood that passes through them? Tick (✓) as many as apply.

empty beaker ☐

conical flask ☐

filter funnel ☐

filter paper ☐

spirit burner ☐

thermometer ☐

ammeter ☐

beaker containing a mixture of rice and peas ☐

7 When a person's kidneys do not work properly due to a disease, they must use a kidney machine to perform the job of the kidneys.

a What is the name of the process carried out by a kidney machine? **Circle** your answer.

- analysis
- catalysis
- dialysis
- filtration
- purification

b Describe how this process is helping the patient.

....

....

....

....

....

....

Photosynthesis

1 Using common names, write a word equation for photosynthesis and include **two** other essential factors above and below the arrow.

2 Van Helmont produced a model of how a plant grows by constructing the equation:

water → mass of plant

This was then used by scientists who followed him. From their investigations, they modified this equation until they created the equation for photosynthesis that we use today.

a Explain why models produced by scientists as a result of investigations can be changed.

....

....

b When a model is produced as the result of scientific investigations today, does it represent a completely accurate scientific explanation? Explain your answer.

....

....

Testing for the carbohydrate called starch in leaves

 3 Akram is testing a leaf for starch.

a What items should he select from this list of equipment? **Circle** your answers.

- beaker
- Liebig condenser
- Bunsen burner
- test tube
- gauze

- conical flask
- microscope
- heat-proof mat
- white tile
- clamp and stand

- filter funnel
- crucible
- tripod

b Akram will need water to perform **three** tasks. What are they? List them in order.

1 ....

2 ....

3 ....

c Akram is using ethanol in his test.

i What does he need it for?

....

ii What safety precautions must he take when using ethanol?

....

....

d Akram completed the first part of the test and the leaf is ready for testing for starch.

i What chemical reagent does Akram need in order to see if starch is present?

....

ii What colour is the reagent in the bottle?

....

iii What colour does the reagent turn if starch is present in the leaf?

....

4 Jamila points out to Akram that he has used a leaf from a plant that has been in a cupboard for two days.

a What result does Jamila predict for Akram's test?

....

b Explain your answer

....

....

Carbon dioxide and starch production

5 Name the **process** which describes what happens to a green plant when it has been placed in a cupboard for several days.

We say the plant has been ....

 6 Budi has set up this plant as the first half of an investigation.

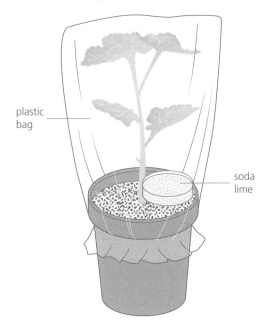

a If the investigation is to work, what must Budi have done to the plant first?

....

....

b What is the purpose of the soda lime?

....

....

c What must Budi do to set up the other half of his investigation?

....

....

d Predict what Budi may find when he makes his starch test at the end of the investigation.

....

....

Light and starch production

7 A plant has been growing in a sunny garden and Asoka tests one of its leaves for starch.

a What will be the result?

....

Joko tidies up the garden and puts a plant pot over the plant by mistake. Two days later, Asoka tests another of its leaves for starch.

b i What will be the result?

....

ii Explain your answer.

....

....

Chlorophyll and starch production

8 Which **two** statements about chloroplasts are **true**. Tick (✓) **two** boxes.

Chloroplasts are found in animal and plant cells. ☐

Chloroplasts contain the green pigment chlorophyll. ☐

Chloroplasts are mainly found in the leaves of a green plant. ☐

All cells in a variegated leaf contain chloroplasts. ☐

Photosynthesis and oxygen

 9 Tahirah sets up an experiment as shown in the diagram. She puts it on a sunny windowsill and leaves it for a week.

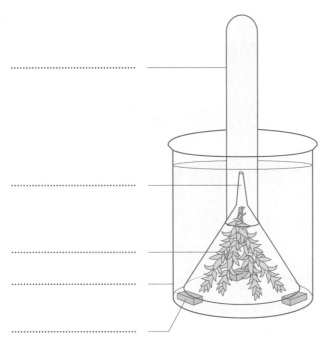

a Label the diagram.

b Describe the appearance of the experiment after one week.

....

....

c **i** What could Tahirah do to take her experiment further?

....

....

ii Predict her result.

....

....

Variation in living things

1

▲ Rabbit

▲ Hare

▲ American desert hare

a Name **one** characteristic that each of these animals has in common.

....

b Give **two** examples of how these creatures are different.

1 ....

2 ....

2 The fins of a fish are shown in the diagram on the right

Labelled A, B, C and D are four fish found in the North Sea on the coast of Europe.

Here is a key to identify the fish.

1	fish with dorsal fin in two parts	*go to 2*
	fish with one long dorsal fin	*go to 3*
2	small anal fin	armed bullhead
	long anal fin	lesser weever
3	large pectoral and pelvic fins	three-bearded rockling
	small pectoral and pelvic fins	lesser sand eel

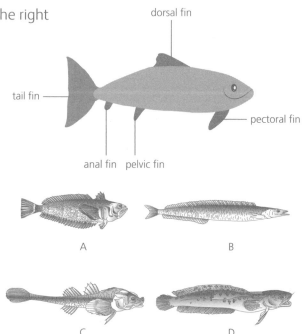

a Use the key to identify each fish.

Fish A is a ....

Fish B is a ....

Fish C is a ....

Fish D is a ....

b One fish is poisonous – it has a tail fin with a straight edge. What is its name?

....

c One fish is the food of many sea birds – it has a forked tail. What is its name?

....

Variation within a species

3 a What is variation within a species?

....

....

b Are all the individuals in a species genetically the same? Explain your answer.

....

....

....

....

....

4 Students investigated variation in the number of peas in a pea pod.

They carefully opened 10 pea pods and counted the number of peas in each pod.

Here are their results.

Number of peas in a pod	2	3	4	5	6	7	8	9
Number of pea pods	1	1	2	3	1	1	0	1

a Complete the frequency chart to show all their results.

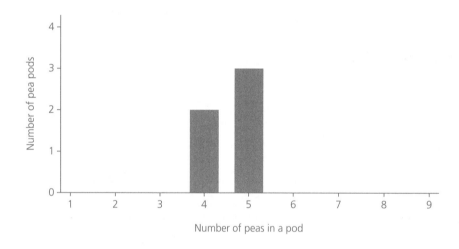

b What number of peas was not found in any of the 10 pods opened?

....

Chromosomes and genes

5 Which **two** of the following statements are correct? Tick (✓) **two** boxes.

There are two genes for each body feature in a cell. ☐

There are two chromosomes for each body feature in a cell. ☐

There is one gene on each chromosome. ☐

There are large numbers of genes on a chromosome. ☐

6 Children (offspring) inherit characteristics from each of their parents. Identical twins inherit the same characteristics, as they divide into two individuals after the fertilised egg forms from the fusion of the male and female gametes.

▲ Identical twins

The table contains information about four different children: A B, C and D.

Use the information to work out who the pair of identical twins are most likely to be.

Characteristic	Child			
	A	B	C	D
Male	✓		✓	
Female		✓		✓
Has lobed ears	✓		✓	✓
Brown hair	✓		✓	✓
Blonde hair		✓		
Brown eyes	✓	✓	✓	

The identical twins are .... and ....

I think this because ....

....

....

Sex chromosomes

7 Show how the gametes of a male and a female can combine at fertilisation using this box and the symbols for the sex chromosomes (X and Y).

Chromosomes and gametes

8 a What is the name for a female reproductive cell. **Circle** your answer.

- chromosome
- egg cell
- gamete

- sperm cell
- zygote

b Complete the sentence.

The male reproductive cell in a human is called a ....

9 The male gamete of an animal has 12 chromosomes.

a How many chromosomes are in the gametes of a female animal of the same species?

....

b These male and female gametes join together. What is this process called?

....

c What is formed as a result of this process and how many chromosomes does it have?

....

....

10 a During the formation of chromosomes in gametes, what happens to the chromosomes?

....

b Why does this chromosome activity affect the appearance of the offspring of the parents?

....

The scientific theory of natural selection

11 In England in the 1850s, most peppered moths were pale-coloured. By 1900, dark-coloured peppered moths were more common.

Underline the possible reasons for this from the sentences below. Underline as many reasons as you think necessary.

- The pale moths changed colour in winter months.
- Pollution from factories landed as soot on tree trunks where peppered moths rest.
- Birds could see pale-coloured peppered moths better and so ate them.
- Dark-coloured peppered moths could fly faster than birds.
- Dark-coloured moths were better camouflaged, so reproduced and more dark-coloured peppered moths were born.

12 a Which statement describes the result of natural selection? Tick (✓) **one** box.

The strongest of two species will survive in a habitat. ☐

The best-suited of two species will survive in a habitat. ☐

The more aggressive of two species will survive in a habitat. ☐

The more numerous of two species will survive in a habitat. ☐

b Which process helps a species change over time to survive in a habitat? Tick (✓) **one** box.

predation ☐

adaptation ☐

conservation ☐

respiration ☐

4 Care in fetal development

Care during fetal development

1 It is important for a pregnant woman to eat healthily during her pregnancy.

Which foods or drinks should a pregnant woman should **avoid**? Tick (✓) as many as apply.

soft cheeses like Camembert ☐

chicken ☐

rice ☐

coffee ☐

fruit ☐

2 Complete the table to show examples of the types of food a pregnant woman should have in her diet during pregnancy.

Food type	Examples
carbohydrate	
protein	
fat	
fruits and vegetables	

3 A pregnant woman's diet is important to help a fetus to grow and develop well.

State **two** things that could happen if a fetus does not grow and develop well.

1 ....

2 ....

4 Nicotine is a harmful substance found in cigarette smoke. The data in the table was collected from a survey of pregnant women about their smoking habits.

Number of cigarettes smoked during pregnancy	0	1–10	More than 10
Number of babies born with low birth weight	2	8	10

    a Use the data to complete the following advice for pregnant women who continue to smoke.

    The .... cigarettes you smoke during pregnancy, the .... the risk of your baby being born with a low birth weight.

    b Compare the information in the table for the three different groups. Are there any limitations in the data that has been collected? Explain your answer.

    ....

    ....

    ....

    ....

    c What could you do to find out more information about the effects of smoking for pregnant women?

    ....

    ....

5 a Name a drug which might be harmful to a fetus if taken by an expectant mother.

    ....

    b Give **two** possible effects it could have on the fetus.

    1 ....

    2 ....

6 a What major problem is common in premature babies?

....

....

b What is the name of the machine used in hospitals that can help premature babies to survive?

....

c Modern versions of this machine control many different things that support the baby. Name three of them.

1 ....

2 ....

3 ....

5 Environmental change and extinction

Organisms in the environment

1 a The water in a habitat collects deep below the soil's surface. How might a species of plant become adapted to reach it?

....

b A species of plant lives in a habitat of large leaf-eating herbivores, such as deer. How might its leaves become adapted to help the species survive?

....

c A species of bird feeds on worms in muddy riverbanks. The worms adapt by burrowing deeper. How might the bird species become adapted so that they can keep feeding on the worms?

....

 2 The table shows the temperature and rainfall records for a 12-month period in one habitat.

Month	Jan	Feb	Mar	Apr	May	Jun	Jul	Aug	Sep	Oct	Nov	Dec
Temperature/°C	37	35	32	28	24	22	22	25	29	32	34	36
Rainfall/mm	45	33	28	12	10	9	6	7	8	20	31	40

a Make a line graph using the data about the temperature.

b Make a bar graph using the data about the rainfall.

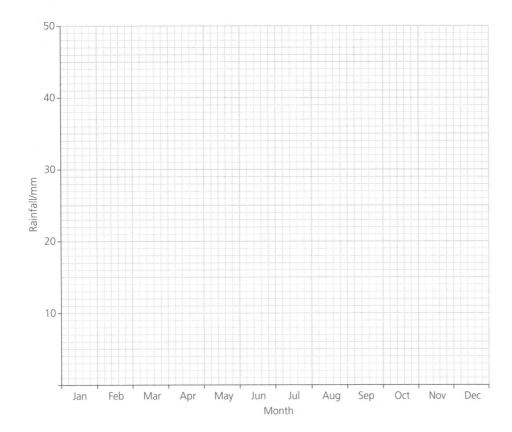

c The data shows that there was a period of very dry weather. Look at your bar graph and decide when you think this period of dry weather began. Write down your answer.

....

d Here are three habitats – rainforest, desert, polar region. Which habitat do you think matches the data in the graphs?

....

e The organisms living in this habitat are adapted to two features of the habitat which occur over the course of a year. Which pair of features are they?

   i High temperatures and high rainfall

   ii Low temperatures and low rainfall

   iii High temperatures and low rainfall.

....

Ecological models: The food web

 3

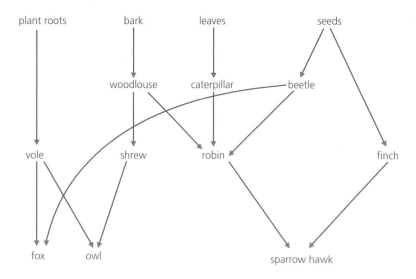

a Name **two** producers in this food web.

1 ....

2 ....

b Name **one** predator.

....

c What do the arrows in a food web represent?

....

....

d Why might a model such as this change over time?

....

....

....

....

Population change

4 A farmer estimated the number of field hares in one of his fields at the same time of year for 10 years. Here are his results.

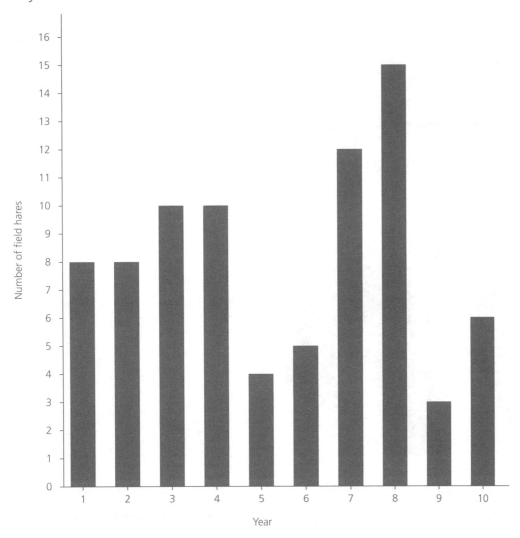

Number of field hares	8	8	10	10	4	5	12	15	3	6
Year	1	2	3	4	5	6	7	8	9	10

a At the end of the eighth year, a footpath was made that runs through the field. What effect did this have on the number of field hares?

....

....

b In another survey on the population of foxes, it was recorded that the population of foxes increased in year 9. How may this information explain the change in the hare population that year?

....

....

....

Endangered species

5 Some animals are under threat of becoming extinct. They are called endangered species.

Some giant pandas live in zoos around the world. How might this prevent them becoming extinct?

....

....

....

....

....

6 In some places around the world, whales are hunted. However, they are not hunted everywhere, because some places have a ban on whale-hunting.

a Give **one** reason why people might choose to hunt whales.

....

....

....

b Is this a good or a bad reason? Explain your answer.

....

....

....

6 The periodic table

Sorting out elements

 1 What are the names Dalton, Döbereiner and Newlands associated with in terms of chemistry?

Dalton...

Döbereiner...

Newlands...

How the periodic table was made

2 What is Mendeleev best remembered for?

....

3 Write a definition for the atomic number of an element.

....

....

4 Look at the diagram of part of the periodic table to answer these questions.

	3	4		5	6	7	8	9	10
Row 2	Li	Be		B	C	N	O	F	Ne
	6.94	9.01		10.8	12.0	14.0	16.0	19.0	20.2
	11	12		13	14	15	16	17	18
Row 3	Na	Mg		Al	Si	P	S	Cl	Ar
	23.0	24.3		27.0	28.1	30.1	32.1	35.5	39.9
	19	20							
Row 4	K	Ca							
	39.1	40.1							

a What is the chemical symbol for potassium? ....

b Which element has the chemical symbol Li? ....

5 Look at the diagram of part of the periodic table. Why are the properties of Be, Mg and Ca similar?

....

....

....

....

6 Which statement describes the atomic number? Tick (✔) **one** box.

the number of electrons in an atom ☐

the number of protons in an atom ☐

the number of neutrons in an atom ☐

the number of protons and neutrons in an atom ☐

Group 1 in the periodic table

7

Element	Melting point/°C	Boiling point/°C
lithium	180	1360
sodium	98	900
potassium	63	

a Which element has the highest melting point? ....

b Describe the trend in melting points as you go down the group.

....

....

c Predict the boiling point of potassium .... °C

8 Draw lines to match each element to the description of its reaction with water.

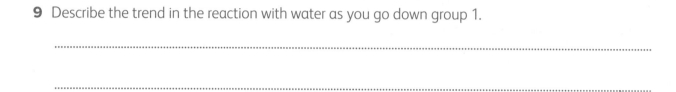

Element	Reaction
lithium	fizzes strongly
sodium	bursts into flame
potassium	fizzes and floats on water

9 Describe the trend in the reaction with water as you go down group 1.

....

....

10 In the table, write down the number of protons, electrons and neutrons of each atom.

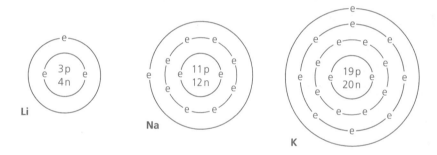

Atom	Number of protons	Number of electrons	Number of neutrons
lithium			
sodium			
potassium			

7 Bonds and structures

The electrons in an atom

1 a Draw lines to match each subatomic particle to its charge.

Subatomic particle		Charge
proton		negative charge
neutron		positive charge
electron		no charge (neutral)

b What is the charge of the atom? **Circle** your answer.

    – positive   – negative   – neutral

2 The diagram shows the structure of a beryllium atom.

    a Complete the labels.

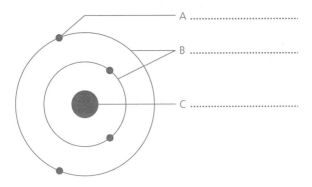

A ....

B ....

C ....

    b i Which part of the atom contains positive charges?

....

    ii Explain your answer.

....

....

3 Draw lines to match the element's name with its symbol. Then decide which atomic structure corresponds to each element. Write a, b, or c in the blank boxes.

    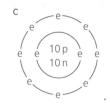

Name	Symbol	Atomic structure
neon	He	....
helium	Be	....
beryllium	Ne	....

Covalent bonding

4 What is the difference between an atom and a molecule?

....

....

5 Sometimes atoms bond with other atoms by sharing electrons.
 – A molecule of water is represented by the formula H_2O.
 – An oxygen atom has six outer electrons.
 – A hydrogen atom has one outer electron.

Draw an atomic diagram showing how a molecule of water is formed from two hydrogen atoms and one oxygen atom. (**Hint:** think about how you might show the covalent bond.)

Ions and ionic bonding

6

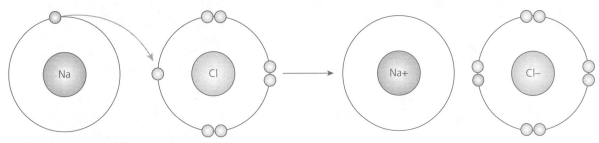

a What is an **ion**?

....

b When sodium combines with chlorine, which is the cation and which is the anion?

Complete the table by ticking (✓) the corresponding element. Then write the charge of the element.

	Sodium	Chlorine
cation		
anion		
charge		

Structures

7 What is an ionic bond?

....

....

8 Complete the table by ticking (✓) **one** box for each statement.

Statement	True	False
Covalent bonds make molecules.		
Most molecules are solids at room temperature.		
Covalent molecules do not usually dissolve in water.		
Covalent bonds do not make molecules.		
Most molecules are liquids or gases at room temperature.		

9

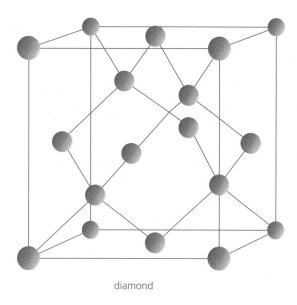

diamond

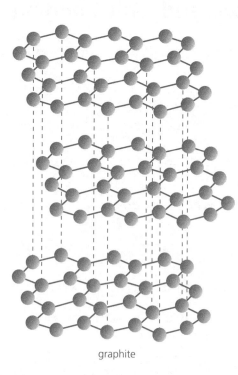

graphite

a How many covalent bonds does each carbon atom form in the structure for diamond?

....

b Describe the structural appearance of graphite, which is a different form of carbon.

....

....

c What unusual property does graphite possess? Can you explain why it has this property?

....

....

....

Giant ionic structures

10 Electrical charges hold these structures together in lattice arrangements.

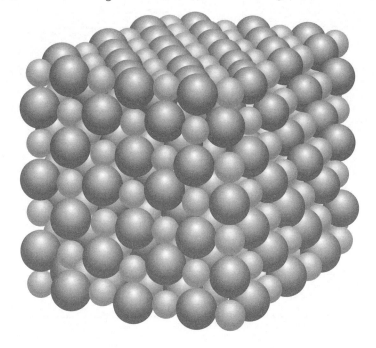

a What is a giant ionic structure made from?

....

....

....

....

b Explain why salt crystals are generally cubic.

....

....

....

....

11 It takes a lot of energy to break the bonds in a crystal lattice.

What can you infer from this about the boiling and melting points of ionic compounds?

....

....

....

12 a What is the formula unit for sodium chloride?

....

b Explain in your own words what a formula unit is.

....

....

Defining and comparing density

1   What is density? Tick (✓) **one** box.

the volume of a substance   ☐

the mass of a substance   ☐

the amount of matter present in a certain volume
of a substance, such as $1\,cm^3$ ☐

volume multiplied by mass   ☐

volume divided by mass   ☐

Density in solids

2   Dembe has three small blocks of different types of wood.

   a   What must he do to compare their densities?

....

   b   Which formula should he use to calculate the density of each wood? Tick (✓) **one** box.

density $=$ mass $\times$ volume   ☐

density $= \dfrac{\text{volume}}{\text{mass}}$ ☐

density $= \dfrac{\text{mass}}{\text{volume}}$ ☐

3   Jaya has a block of material and wants to find its volume.

   a   How should she do this?

....

....

b Jaya's block is a cube with sides of 5 cm. What is its volume? Show your working.

....

....

....

4 Shazad wants to find the density of a pebble.

a Here are the stages he should use but they are in the wrong order. Arrange them correctly by numbering them in the order in which they occur. The first statement has been done for you.

A Put the pebble on a balance. ☐ 1

B Attach a string to the pebble. ☐

C Half-fill a measuring cylinder with water. ☐

D Read the volume of water and pebble. ☐

E Carefully lower the pebble into the measuring cylinder. ☐

F Make sure that the pebble is completely covered by water. ☐

G Read off mass of pebble. ☐

H Read the first level of water. ☐

b How should Shazad find the volume of the pebble?

....

....

c Shazad's pebble has a mass of 90 g and a volume of 30 cm³. What is its density? Show your working.

....

....

....

Density in liquids

5 Fozia wants to find the density of honey.

a Here are the stages she should use but they are in the wrong order. Arrange them correctly by numbering them in the order in which they occur. The first statement has been done for you.

A Read the mass of the measuring cylinder (A). ☐

B Read the mass of the cylinder and liquid (B). ☐

C Read the volume of liquid in the measuring cylinder (V). ☐

D Put the measuring cylinder on the balance. 1

E Place the cylinder containing the liquid on the balance. ☐

F Pour the liquid into the measuring cylinder. ☐

b Use the following letters: A = mass of measuring cylinder, B = mass of honey and measuring cylinder, V = volume of honey in measuring cylinder. Which formula should Fozia use to calculate the density of the honey? Tick (✔) **one** box.

$$density = \frac{A - B}{V}$$ ☐ $$density = \frac{A + B}{V}$$ ☐

$$density = \frac{B - A}{V}$$ ☐ $$density = \frac{V - A}{B}$$ ☐

c Fozia's results are A = 100 g, B = 120 g and V = 14 cm³. What is the density of the honey? Show your working.

....

....

....

Floating and sinking

6 Budi has three liquids: water, vegetable oil and maple syrup. He says the water has a density of 1, the vegetable oil has a density of 0.92 and the maple syrup has a density of 1.37.

a Using what you know about the density of water, what are the units he is using to measure the density?

....

b What would the units be if he multiplied them by one thousand?

....

c Budi pours each one into the same tall jar. They settle out in layers.

i Which liquid is at the top?

....

ii Which liquid is in the middle?

....

iii Which liquid is at the bottom?

....

iv Explain your answers.

....

....

....

....

d Budi pours some corn oil into the jar. He says it has a density of 0.97. Where does it settle? Tick (✔) **two** boxes.

above the water	☐	above the maple syrup	☐
below the vegetable oil	☐	below the maple syrup	☐

Density in gases

7 Explain how to calculate the density of a gas. Use the diagram below to help you.

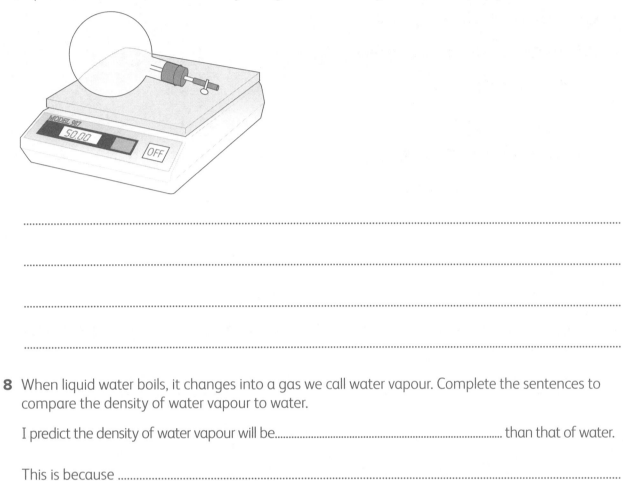

....

....

....

....

8 When liquid water boils, it changes into a gas we call water vapour. Complete the sentences to compare the density of water vapour to water.

I predict the density of water vapour will be... than that of water.

This is because ....

....

9 A teacher is demonstrating how to find the density of air. Here are the stages she uses but they are in the wrong order. Arrange them correctly by numbering them in the order in which they occur. The first statement has been done for you.

A Put the evacuated flask on the balance. ☐

B Pour water from the flask into a measuring cylinder. ☐

C Find the mass of the flask. ☐

D Open the flask underwater. ☐

E Measure the volume of the water. ☐

F Place the flask on the balance. ☐

G Let the water replace the vacuum in the flask. ☐

H Measure the mass of the evacuated flask. ☐

I Remove the air from the flask with a vacuum pump. 1

10 a The particles of a gas move further apart when they are heated. Describe how this affects the density of the gas.

....

....

....

....

b The particles of a gas are pushed closer together when the pressure on the gas is increased. Describe how this affects the density of the gas.

....

....

....

....

c If the densities of two gases are measured at the standard temperature and pressure (STP), what does it mean?

....

....

....

....

11 a Which gas is now used in gas-filled balloons?

....

b Describe one way in which gas-filled balloons are used for scientific research.

....

....

9 | Displacement reactions

Displacement reactions

1 a If an iron nail were placed in the copper sulfate solution, would you see a change? **Circle** your answer.

- Yes
- No

b Explain your answer.

....

....

2 When a copper wire is placed in a solution of silver sulfate, the solution turns blue and silver appears on the wire.

a Why does the solution turn blue?

....

....

b Where does the silver come from?

....

....

c Why does this change take place?

....

....

3 What is the word equation for the following displacement reaction?

$Mg + H_2O \rightarrow MgOH_2 + H_2$

....

....

The reactivity series of metals

4 Write a sentence to describe what happens when each metal is heated. You should use one or more of the following words and phrases in each of your sentences.

- black powder
- glows
- not changed

- makes yellow sparks
- forms black powder on the surface

a iron

....

b gold

....

c copper

....

5 What metal am I? Read the clues to identify the metal.
- I burn very brightly in air to form a metal oxide.
- I react slowly in cold water to form a hydroxide.
- I react vigorously in dilute acid.

Circle your answer.

- aluminium   - calcium   - copper   - gold   - potassium

6 Here are some metals from the reactivity series.

List the top **five** in the correct order.

- aluminium
- calcium
- copper
- gold
- iron

- magnesium
- potassium
- sodium
- tin
- zinc

1 ....

2 ....

3 ....

4 ....

5 ....

The reaction of metals with acids

7 Which word equation shows the correct reaction for a metal reacting with an acid? Tick (✔) **one** box.

metal + acid → metal oxide + hydrogen ☐

metal + acid → metal salt + water ☐

metal + acid → metal salt + oxygen ☐

metal + acid → metal salt + hydrogen ☐

metal + acid → metal oxide + water ☐

8 A class carried out the reaction of some different metals with dilute acids. The results are shown in the table.

Metal	Time taken to collect one full boiling tube/seconds
Mg	20
Zn	56
Fe	65

a Describe the trend in these results.

....

....

b What product was being collected in the boiling tube?

....

9 Jaafar is investigating the reaction of some metals with an acid.

a After he has assembled his apparatus, put in the metal and added the acid, there is a fizzing and a gas is produced. What is this gas? ....

b Jaafar has used hydrochloric acid in his investigation and has begun to write the word equation. Complete the word equation for him.

metal + hydrochloric acid → ....

c He decides to set up the apparatus again and try a second metal.

How could he find out whether the metal is more reactive or less reactive than the one he has just investigated?

....

....

....

....

d If Jaafar tested the following metals, which ones would he see react with the acid? **Circle** your answers.

- calcium
- copper
- gold
- magnesium
- silver
- zinc

Preparing common salts

Salts

1 Identify **two** uses of calcium chloride. Tick (✓) **two** boxes.

herbicides ☐   setting concrete ☐

drying agents ☐   bleaching paper ☐

2 Identify **two** uses of zinc sulfate. Tick (✓) **two** boxes.

making cosmetics ☐   sewage treatment ☐

food processing ☐

Acids and their salts

3 Name the salt produced by each acid.

hydrochloric acid ....

sulfuric acid ....

nitric acid ....

4 Write the word equations for the following reactions:

a magnesium and nitric acid ....

b lead and hydrochloric acid ....

c aluminium and sulfuric acid ....

5 a How do the fibres in a filter paper help in the process of filtration?

....

....

b What is the filtrate?

....

....

c What is the residue?

....

....

6 Describe what happens in the process known as evaporation.

....

....

....

7 a What is a crystal?

....

....

b How would you make crystals of a substance?

....

....

....

Preparing a salt from a metal and an acid

8 a Label the diagram for the preparation of zinc chloride below.

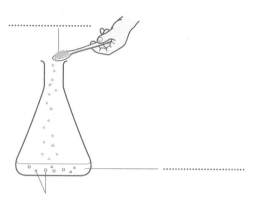

b These are the stages involved in the preparation of zinc chloride, but they are in the wrong order. Arrange them correctly by numbering them in the order in which they occur.

    A The flask contents are poured through a filter paper in a filter funnel. ☐

    B The reactants are mixed together in a conical flask. ☐

    C Metal remains on the filter paper and liquid passes through into a beaker. ☐

    D Bubbles are formed and pass through the acid. ☐

c By what process is the final product obtained?

....

Preparing a salt from a metal carbonate and an acid

9 a These are the stages in the preparation of a salt from a metal carbonate, but they are in the wrong order. Arrange them correctly by numbering them in the order in which they occur.

    A Look for the end of bubbling in the mixture. ☐

    B Add the solid to the acid. ☐

    C Leave the heated solid to cool. ☐

    D Heat the liquid in a heat-proof container until a solid appears. ☐

    E Pour the acid into a container. ☐

    F Transfer the separated liquid to a heat-proof container. ☐

    G Separate the solid from the liquid. ☐

b What process is performed at the end to remove any remaining liquid?

....

10 a Write the equation for the preparation of copper chloride from copper carbonate and hydrochloric acid.

....

....

b Why do you need to add excess copper carbonate?

....

c Describe **one** safety precaution you would need to take if you carried out this reaction.

....

11 Hans is preparing copper sulfate using sulfuric acid. He has a sample of copper metal and a sample of copper carbonate.

a Which sample should he use?

....

....

b Explain your answer.

....

....

Rates of reaction

Rates of reaction

1 What does the term 'rate' mean? Tick (✓) **one** box.

a measure of the rise of temperature in a reaction ☐

a measure of the amount of chemicals in a reaction ☐

a measure of the speed of change in a reaction ☐

a measure of the concentration of the reactants ☐

2 In every chemical reaction, the mass of the matter is conserved. What does this mean?

....

....

....

....

3 Which of these statements is **true** about the mass of the reactants and products in a chemical reaction? Tick (✓) **one** box.

The mass of the reactants is the same as the mass of the products. ☐

The mass of the reactants is greater than the mass of the products. ☐

The mass of the reactants is smaller than the mass of the products. ☐

When a chemical reaction takes place, the mass of the products gets larger and larger. ☐

Measuring rates of reaction

4 Sarah is planning to measure the change in mass during a reaction.

a What piece of equipment does she need to do this?

....

b Here are Sarah's instructions but they are in the wrong order. Arrange them correctly by numbering them in the order in which they occur.

A Find the mass of the mixed reactants. ☐

B Plot the data collected on a graph. ☐

C Find the mass of the separate reactants. ☐

D Record the mass of the reactants frequently over a period of time. ☐

E Put the reactants together. ☐

c After her first investigation, Sarah decides she will measure the volume of a gas produced in a reaction. What special piece of equipment will she need for this investigation?

....

5 a A quick way to measure the rate of a reaction is by measuring how quickly any gas is produced. What piece of equipment would you use to do this?

....

b A class carried out an investigation into the rate of a reaction. The results are shown in the table below.

Time/seconds	Volume of gas produced/cm³
0	0
30	8
60	18
90	24
120	38
150	44
180	50
210	56
240	60
270	60

How long did the reaction last? .... seconds

c Explain how you know.

....

....

d Draw a graph of the results in the space below.

e Estimate how much gas might have been produced after 45 seconds.

....

f Which is the anomalous result? **Circle** it on your graph.

g What might be the reason for this anomalous result?

....

....

Factors affecting rates of reaction

6 What is the concentration of a liquid? Tick (✓) **one** box.

a measure of the amount of solvent in the solution ☐

a measure of the amount of solute in the solution ☐

a measure of the amount of solvent in the solute ☐

a measure of the amount of solution in the solvent ☐

7 a Which type of reactant would you choose to obtain the most vigorous reaction? Tick (✓) **one** box.

a single solid lump ☐

smaller lumps ☐

a powdered form ☐

b Explain your reasoning. ....

....

8 Shazia takes a lump of reactant and adds it to another substance and records the time when the reaction stops. It takes 5 minutes. She then grinds up another lump of reactant, the same size as the first, with a pestle and mortar.

a What is happening to the size of the particles she makes as she grinds the reactant?

....

b What is happening to the surface area of the reactant as she grinds it down?

....

c i If Shazia repeats the experiment with the powder, how long will the reaction take? Tick (✓) **one** box.

more than 5 minutes ☐

5 minutes ☐

less than 5 minutes ☐

ii Explain your answer.

....

....

9 a How can the rate of reaction be measured when sodium thiosulfate reacts with hydrochloric acid?

....

b How will you know when to stop timing?

....

c The table shows the reaction times at different temperatures.

Temp/°C	Time/seconds
20	100
25	70
30	50
35	36
40	28
45	20

Explain in your own words how temperature affects the rate of reaction.

....

....

....

....

10 David is investigating the volume of gas produced in a reaction. Here are his results.

Time/minutes	Volume/cm³
0	0
1	6
2	12
3	18
4	22
5	24
6	25
7	25

a Plot the data on the graph below.

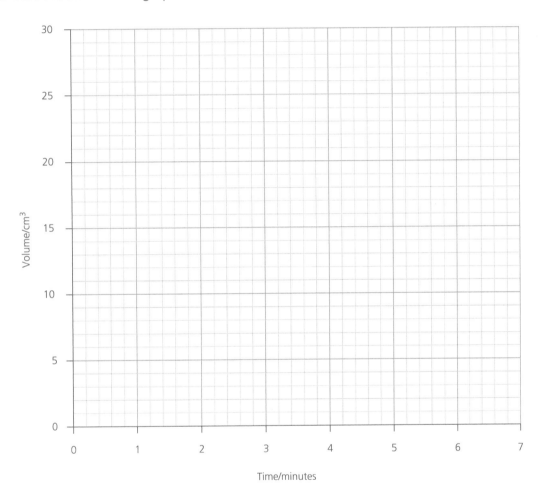

Time/minutes

The laboratory temperature was 25 °C. The teacher switches on the air conditioning to make it cooler. David repeats the investigation and plots a second graph.

b i Where will the line of his second graph be? Tick (✓) **one** box.

at the same place as the first graph ☐

above the line of the first graph ☐

below the line of the first graph ☐

ii Explain your answer.

....

....

12 Energy

Thermal energy, internal energy and temperature

1 What is heat a measure of? Tick (✓) **two** boxes.

the total kinetic energy of atoms and molecules in a substance ☐

part of the kinetic energy of atoms and molecules in a substance ☐

the potential gravitational energy of atoms and molecules in a substance ☐

the total movement energy of atoms and molecules in a substance ☐

2 What is a temperature? Tick (✓) **one** box.

an indication of how hot an object is ☐

an indication of how cold an object is ☐

an indication of how hot or cold an object is ☐

an indication of how warm an object is ☐

3 A pan of water containing 50 cm³ of water is heated on a stove.

a Would this water boil faster or slower than 100 cm³ water on the same stove? **Circle** your answer.

   – faster
   – slower

b What is this due to? **Circle** your answer.

   – density
   – mass
   – substance
   – temperature
   – volume

Measuring the amount of thermal energy

4 Rafiq is finding out how much energy is in a quantity of candle wax. He sets up the apparatus shown in the diagram.

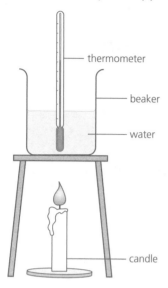

He records the mass of the candle on a balance and records the temperature of the 500 cm³ of water in the beaker.

He lights the candle, stirs the water and records the water temperature until it has risen 10 °C.

He puts out the burning candle and records its mass again.

Here are his results.

Mass of candle before burning	10 g
Mass of candle after burning	5 g
Loss of mass	_____ g

Temperature before heating	22 °C
Temperature after heating	32 °C
Rise in temperature	_____ °C

a Complete the two tables.

b The energy in 1 g of candle is found by using the following formula:

$$\frac{2.1 \times \text{rise in temperature}}{\text{loss of mass}} = \text{kJ/g}$$

Use the formula to find the energy in the candle wax that was burnt away by Rafiq. Show your working.

....

....

....

5 Jaya is comparing the heat produced by two fuels. She uses each one in turn to heat the same volume of water for 10 minutes. Then she records the temperature every 2 minutes.

The table shows her data.

Time/minutes	Fuel A temperature/°C	Fuel B temperature/°C
0	20	20
2	22	24
4	24	28
6	26	32
8	28	36
10	30	40

a Draw and label one line graph with a separate line for each fuel in the space below.

b Use your graph to predict what the temperature of the water would be after 12 minutes. Mark your prediction with a dot on the graph.

c Predict the temperature difference between the two beakers of water after 12 minutes.

Conservation of energy

6 Explain in your own words what the phrase 'conservation of energy' means.

....

....

....

Heat dissipation

7 Which direction does heat (thermal) energy always travel in? Tick (✓) **one** box.

cold to hot ☐   it doesn't travel ☐

from hotter to cooler ☐   cold to warmer ☐

cooler to hotter ☐

8 One end of the metal rod is heated using a Bunsen burner or heat source.

The metal thumb tacks (drawing pins) are attached to the metal rod with wax.

a Predict what will happen as the metal rod is heated.

....

b Describe how this equipment could be used to compare the conductivity of different metals.

....

....

....

9 Explain how convection occurs when water is heated in a saucepan.

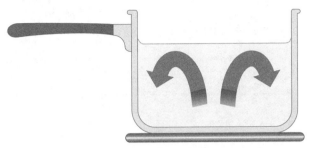

Use these terms to help you:

- convection current
- heat energy
- kinetic energy
- particles
- convection

....

....

....

....

....

10 a Describe how energy travels as radiation.

....

....

....

b Energy can travel through air, but what else can it travel through?

....

Evaporation

11 a On a warm day, after a rain shower, puddles of water remain on the ground. Explain the effect evaporation will have on the puddles.

....

....

b What state is the water in when it 'disappears'? **Circle** your answer.

– gas   – liquid   – solid

c Describe another example of this process happening in everyday life.

....

....

....

d Do things become hotter or cooler when this process happens? Explain your answer.

....

....

....

12 a After vigorous exercise, we produce sweat on the skin surface. When we stop, the sweat evaporates. What effect does this have on body temperature?

....

b Explain why this happens.

....

....

....

13 Waves

Sound and vibrations

1 What happens when you 'twang' a ruler and move one end over the edge of a table and back? Tick (✓) **two** boxes.

The sound goes higher as the vibrating length increases. ☐

The sound goes lower as the vibrating length increases. ☐

The sound goes higher as the vibrating length decreases. ☐

The sound goes lower as the vibrating length decreases. ☐

2 Explain in your own words what a vibration is.

....

....

....

....

3 a When something vibrates in the air, what get squashed together?

....

....

b What change in the air does this squashing cause? ....

....

c When the squashing ends, what happens next?...

....

d What alternating regions are generated in the air?

....

....

e What do these alternating regions make as they move through the air?

....

....

The features of a waveform

4 In the waveform for the sound wave shown below, what name is given to both the distances labelled A and both the distances labelled B?

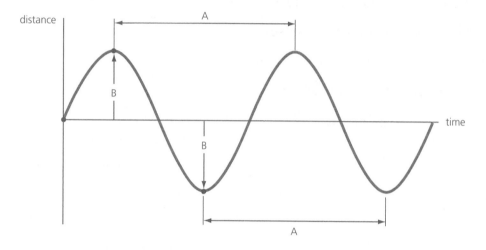

A ....

B ....

5 Which of the waveforms represents the louder sound? **Circle** your answer.

a

b

6 Which statement best describes pitch? Tick (✓) **one** box.

Pitch is how loud a sound is. ☐

Pitch is how high or low a musical note sounds. ☐

Pitch is how soft a musical note is. ☐

Pitch is how long a sound can be heard. ☐

Pitch is how many sounds can be heard in one minute. ☐

7 a What is frequency? ....

....

b Label diagrams A and B as being either a high-frequency sound or low-frequency sound.

a

A ....

b

B ....

c What **units** do we use to measure the frequency of a sound? Give its name and symbol.

Name: ....

Symbol: ....

When sound waves interact

8 a What is the term used to describe when two sounds waves meet? Tick (✓) **one** box.

interfere ☐   frequency ☐

resonate ☐   pitch ☐

amplitude ☐

b Give **two** words to define each description below.

i When two sound waves of the same frequency meet, producing a wave of greater amplitude.

.... and ....

ii When two sound waves of the same frequency meet and produce a wave of smaller amplitude.

.... and ....

c Explain in your own words what a 'dead spot' is.

....

....

....

Modelling sound waves

9 a What properties of waves are measured when using a ripple tank? Tick (✓) as many as apply.

frequency ☐   wavelength ☐   speed ☐

amplitude ☐   current ☐

b The ripple tank is used to model sound waves. Describe the strengths and limitations of this model.

Strengths: ....

....

....

Limitations: ....

....

....

Circuits

1 Naveen sets up the circuit shown in the diagram.

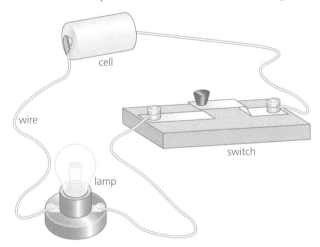

a When he closes the switch, which of the following happens in the wire? Tick (✓) **one** box.

The electrons move from the negative terminal to the positive terminal. ☐

The atoms move from the negative terminal to the positive terminal. ☐

The electrons move from the positive terminal to the negative terminal. ☐

The atoms move from the positive terminal to the negative terminal. ☐

b Where does the energy to create the current come from? **Circle** your answer.

– wire   – cell   – lamp   – switch

c As the current flows through the circuit, the wire in the lamp behaves differently from the other wires in the circuit. What happens to the wire?

....

Naveen leaves the circuit on for some time and later observes that the current has become weaker.

d How can he tell just by looking at the circuit?

....

e What has caused this change in the current?

....

f Naveen opens the switch.

   i What happens to the current?

....

   ii Explain your answer.

....

....

2 Here are the symbols for the components in Naveen's circuit. The one on the left is for the cell.

a Put a cross (×) next to the side which has the negative terminal.

b Make a circuit diagram of Naveen's circuit using these symbols.

c Naveen adds another cell in series to the circuit. Draw the circuit diagram for this new circuit.

d Naveen switches on the circuit.

i How does the appearance of the lamp change?

....

ii Explain your answer.

....

....

3 The connection of the wire to the lamp breaks. Aruni suggests putting a piece of aluminium foil across the gap. Naveen disagrees and says a piece of wood is all that is needed.

a Who is correct? ....

b Explain your answer.

....

....

4 Here are the symbols of three components of circuits. Name them.

A   B   C

A ....

B ....

C ....

5 An electric current is flowing through a wire, then the wire splits into two to form a parallel circuit. What happens to the current? Tick (✓) **one** box.

It stops. ☐

All of it flows down one wire and none of it flows through the other wire. ☐

It splits up and flows equally through both wires. ☐

It splits up but more of the current flows through one wire than the other. ☐

Measuring current

6 Which piece of equipment is used to measure current in a circuit? **Circle** your answer.

   – ammeter   – buzzer   – cell   – lamp   – voltmeter

7 An ammeter has its positive terminal marked in red.

When adding an ammeter to a circuit, which instruction should you follow? Tick (✓) **one** box.

Connect the red terminal to the positive terminal of the cell. ☐

Connect the red terminal to the negative terminal of the cell. ☐

Connect the red terminal to either terminal of the cell. ☐

Connect the red terminal to a lamp. ☐

8 Ochi sets up the circuit shown in the diagram. She connects the ammeter at point A and then at point B.

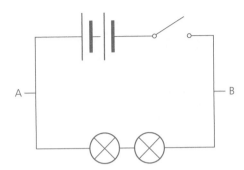

What does she find? Tick (✓) **one** box.

The reading at A is lower than at B. ☐

The reading at A is higher than at B. ☐

The reading at A is the same as at B. ☐

9 Kali sets up the circuit shown in the diagram.

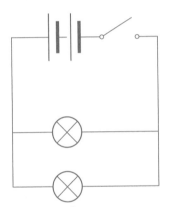

a How many places in the circuit should she check with the ammeter? **Circle** your answer.

- 2
- 3
- 4
- 5

b Mark the places on the circuit diagram.

Measuring voltage

10 What does voltage measure? Tick (✓) **one** box.

a difference in potential energy ☐

a difference in current speed ☐

a difference in resistance ☐

a difference in lamp brightness ☐

11 The diagram shows the symbol for the voltmeter and a simple circuit.

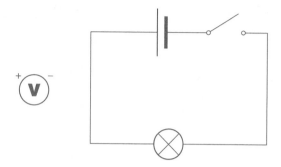

Draw how you would connect the voltmeter to the circuit to measure the voltage across the lamp.

Resistance

 12 a What unit is used to measure resistance? Give its name and symbol.

Name: ....

Symbol: ....

b What is the equation that is used to calculate resistance? Write it in both word and symbol form.

Words: ....

Symbols: ....

c What is the resistance of a lamp when the voltage across it is 12 V and the current is measured as 3 A?

....

....

d If the lamp is replaced with a different component that has a resistance of 6 Ω and the voltage remains 12 V, what will be the new current?

....

....

15 Planet Earth

Evidence for tectonic plates

1 Explain in your own words what a tectonic plate is.

....

....

....

....

2 Scientists have classified rocks into three groups. What are they?

1 ....

2 ....

3 ....

3 a Describe how most fossils form.

....

....

....

....

b Fossils such as these have been found on different continents. What explanation can account for this?

....

....

....

....

4 What pattern links the locations where volcanoes and earthquakes occur around the world with the tectonic plates? Tick (✓) **one** box.

There is no pattern. ☐

Volcanoes and earthquakes occur often at plate boundaries. ☐

Volcanoes and earthquakes occur often away from plate boundaries. ☐

Volcanoes and earthquakes occur in different places. ☐

5 a What is the name of the molten rock that pushes up between the tectonic plates in the ocean?

....

b What happens to this rock when it meets the cold ocean water? Mention **two** processes.

1 ....

2 ....

c Explain in your own words what happens to the minerals in the rock when this change takes place.

....

....

....

d Explain how scientists can use this evidence in their study of tectonic plates.

....

....

....

How tectonic plates move

6 What do seismometers do? Tick (✓) **one** box.

measure the depth of the oceans ☐

measure the strength of volcanoes ☐

measure the strength of earthquakes ☐

measure the depth of underwater trenches ☐

7 Name the four major components of the structure of the Earth. Start with the innermost and move closer to the surface.

1 ....

2 ....

3 ....

4 ....

8 a What is generated at the centre of the Earth and moves outwards towards its surface, causing the mantle to move?

....

b What kinds of currents occur in the mantle?

....

c Explain in your own words how these currents affect the tectonic plates.

....

....

....

16 Cycles on Earth

Processes in the carbon cycle

1 a How do plants receive carbon dioxide? Tick (✓) **one** box.

by their roots ☐

from the air ☐

when they are eaten ☐

in water ☐

room sunlight ☐

b Write the word equation for photosynthesis.

....

c Where do plants get their water from?

....

d What substance is the product of photosynthesis and is stored in the leaves of plants?

....

2 The word equation for the process of respiration is

glucose + oxygen → carbon dioxide + water

What is the energy produced from respiration used for in the human body?

....

....

3 a Insert arrows in the food chain below.

b Name the secondary consumer in this food chain.

....

4 a In your own words, explain the similarities and differences between a food chain and a food web.

....

....

....

....

b Do you think a food web is a more useful model than a food chain? What strengths or limitations do the two models have?

....

Strengths:...

....

....

Limitations:...

....

....

5 What do decomposers release during respiration?

....

6 a Give **two** uses of combustion early humans found useful in their daily lives.

1 ....

2 ....

b Write the word equation for combustion.

....

7 The carbon cycle shows how carbon is recycled in the environment.

Give **two** ways that carbon dioxide is released into the air.

1 ....

2 ....

8 In the space below, draw your own diagram of the carbon cycle.

Carbon dioxide the 'greenhouse gas'

9 List **two** things that can increase the amount of carbon dioxide in the air.

1 ....

2 ....

10 What does it mean when we say that 'carbon dioxide is a greenhouse gas'?

....

....

11 Why is the natural greenhouse effect important for living things on Earth?

....

....

Climate change

12 Over the past 100 years, carbon dioxide has been entering the atmosphere more quickly than it is being removed, leading to climate change.

Give **one** example of a sign of climate change.

....

13 Name three features of the atmosphere that can be measured at weather stations.

1 ....

2 ....

3 ....

14 a Why did the Industrial Revolution lead to a greater demand for energy?

....

....

b What fuel was originally used to provide energy through combustion?

....

c How did the use of this fuel affect the atmosphere?

....

....

d Later, another fuel was used. What was it?

....

The effects of climate change

15 a In what major way is the climate of the Earth specifically predicted to change and what is this change called?

....

....

b How are sea levels predicted to change? Explain why.

....

....

....

....

c Describe some of the possible effects of sea level change.

....

....

....

....

16 a How will the water content of the atmosphere change? Explain why.

....

....

....

....

b How will the change in the water content in the atmosphere affect the flow of air around the planet? What effects might this have?

....

....

....

....

c In some places, there will be an increase in rainfall. What may this cause on the ground?

....

d In some places there will be a great decrease in rainfall. What may this cause on the ground?

....

17 a Name three types of extreme weather events.

1 ....

2 ....

3 ....

b How does climate change contribute to hurricanes causing more damage?

....

....

....

....

c What other effects do scientists think climate change may have on extreme weather events?

....

....

....

....

Earth in space

Asteroids

1 Between which two planets is the asteroid belt found? Tick (✓) **one** box.

Earth and Mars ☐   Saturn and Uranus ☐

Uranus and Neptune ☐   Mars and Jupiter ☐

Mercury and Venus ☐

2 a Write a sentence to explain what holds asteroids in the asteroid belt in position.

....

....

b What does an asteroid cause on impact when it collides with a planet or a moon?

....

c Where in the solar system can we see clear examples of the feature you have named in Question 2b, which remain exactly as they appeared when they were first formed?

....

d Explain in your own words why the example in the place you have named in Question 2c remain exactly as they were when they first formed.

....

....

....

....

3 a List the materials you could use to model an asteroid collision.

....

....

b Explain how you would use the materials to simulate an asteroid collision.

....

....

....

....

c What relationship could be investigated with this model?

....

....

d Describe the strengths and limitations of using this model.

Strengths:...

....

....

Limitations:..

....

....

The effects of asteroid collision on Earth

4 a What are the effects on Earth's atmosphere of an asteroid that hits a land mass?

....

b What are the effects on Earth's atmosphere of an asteroid that hits a body of water?

....

c What effect do these both cause?

....

5 Name two conditions of an impact winter.

1 ....

2 ....

6 What is the most serious possible effect an impact winter can have on animals and plants on Earth? **Circle** your answer.

- increased human activity
- lack of food
- increase in volcanic eruptions
- planetary tilt

The formation of the Moon

7 a What is the name of the collision theory for how the Moon was formed?

....

b If Moon rock samples show evidence of having been molten, what does that indicate?

....

....

Where stars are born

8 What are the clouds of dust and gas called where it is believed stars are formed?

....

9 What name is given to the place where stars are formed?

....

10 In your own words, describe and explain how a star is formed.

....

....

....

....

11 A model of a stellar nursery was made in the following way.
  - A translucent material, such as muslin, was held vertically like a curtain.
  - Behind it, a lamp was connected to a cell and switch and the current was allowed to flow.
  - A photograph was taken of the shining lamp.
  - Another cell was added in series and the lamp was photographed again.
  - Finally, a third cell was added in series and another photograph taken.

a What was the translucent material modelling?

....

b How did the shining of the lamp change as more cells were added to the circuit?

....

c What are the strengths and limitations of this model in demonstrating how stars are born?

Strengths:...

....

....

Limitations:..

....

....

The **Cambridge Checkpoint Lower Secondary Science** series consists of a Student's Book, Boost eBook, Workbook and Teacher's Guide with Boost Subscription for each stage.

Student's Book	Boost eBook	Workbook	Teacher's Guide with Boost subscription
Student's Book 7 9781398300187	eBook 7 9781398302136	Workbook 7 9781398301399	Teacher's Guide 7 9781398300750
Student's Book 8 9781398302099	eBook 8 9781398302174	Workbook 8 9781398301412	Teacher's Guide 8 9781398300767
Student's Book 9 9781398302181	eBook 9 9781398302228	Workbook 9 9781398301436	Teacher's Guide 9 9781398300774

The audio files are FREE to download from:
www.hoddereducation.com/cambridgeextras

To explore the entire series,
visit **www.hoddereducation.com/cambridge-checkpoint-science**

Cambridge Checkpoint Lower Secondary Science Teacher's Guide with Boost subscription

Created with teachers and students in schools across the globe, Boost is the next generation in digital learning for schools, bringing quality content and new technology together in one interactive website.

The **Cambridge Checkpoint Lower Secondary Science Teacher's Guides** include a print handbook and a subscription to Boost, where you will find a range of online resources to support your teaching.

- **Confidently deliver the new curriculum framework:** Expert guidance on the different approaches to learning, including developing scientific language and the skills required to think and work scientifically.

- **Develop key concepts and skills:** Suggested activities, knowledge tests and guidance on assessment, as well as ideas for supporting and extending students working at different levels.

- **Support the use of ESL:** Introductions and activities included that have been developed by an ESL specialist to help facilitate the most effective teaching in classrooms with mixed English abilities.

- **Enrich learning:** Audio versions of the glossary to help aid understanding, pronunciation and critical appreciation.

To purchase Cambridge Checkpoint Lower Secondary Science Teacher's Guide with Boost subscription, visit www.hoddereducation.com/cambridge-checkpoint-science

Cambridge checkpoint

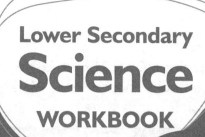

Lower Secondary Science WORKBOOK

9

Practise and consolidate knowledge gained from the Student's Book with this, write-in workbook full of corresponding learning activities.

- Save time when planning with ready-made homework or extension exercises.
- Reinforce students' understanding of key scientific concepts with varied question types and the use of ICT.
- Challenge students with extra practice activities to encourage regular self-assessment.

For over 30 years we have been trusted by Cambridge schools around the world to provide quality support for teaching and learning. For this reason we have been selected by Cambridge Assessment International Education as an official publisher of endorsed material for their syllabuses.

Working for over **30 YEARS** *WITH Cambridge Assessment International Education*

For more information on how to use this workbook, please visit: **www.hoddereducation.com/workbook-info**

This resource is endorsed by Cambridge Assessment International Education

✓ Provides learner support as part of a set of resources for the Cambridge Lower Secondary Science curriculum framework (0893) from 2020

✓ Has passed Cambridge International's rigorous quality-assurance process

✓ Developed by subject experts

✓ For Cambridge schools worldwide

Boost
This series includes eBooks and teacher support.
Visit www.hoddereducation.com/boost for more information.

Registered Cambridge International Schools benefit from high-quality programmes, assessments and a wide range of support so that teachers can effectively deliver Cambridge Lower Secondary.

Visit **www.cambridgeinternational.org/lowersecondary** to find out more.

HODDER EDUCATION
e: education@hachette.co.uk
w: hoddereducation.com

ISBN 978-1-398-30143-6

9 781398 301436